Tucholsky Wagner Zola Scott Sydow Fonatne Freud Schlegel
Turgenev Wallace
Twain Walther von der Vogelweide Fouqué Friedrich II. von Preußen
Weber Freiligrath Frey
Fechner Fichte Weiße Rose von Fallersleben Kant Ernst Richthofen Frommel
Hölderlin
Engels Fielding Eichendorff Tacitus Dumas
Fehrs Faber Flaubert
Eliasberg Ebner Eschenbach
Feuerbach Maximilian I. von Habsburg Fock Eliot Zweig
Ewald Vergil
Goethe London
Elisabeth von Österreich
Mendelssohn Balzac Shakespeare
Lichtenberg Rathenau Dostojewski Ganghofer
Trackl Stevenson Doyle Gjellerup
Mommsen Tolstoi Hambruch
Thoma Lenz Hanrieder Droste-Hülshoff
Dach Verne von Arnim Hägele Hauff Humboldt
Reuter
Karrillon Garschin Rousseau Hagen Hauptmann Gautier
Damaschke Defoe Hebbel Baudelaire
Descartes
Wolfram von Eschenbach Schopenhauer Hegel Kussmaul Herder
Bronner Darwin Melville Dickens Grimm Jerome Rilke George
Campe Horváth Aristoteles Bebel Proust
Bismarck Vigny Gengenbach Barlach Voltaire Federer Herodot
Heine
Storm Casanova Lessing Tersteegen Gilm Grillparzer Georgy
Chamberlain Langbein Gryphius
Brentano Lafontaine
Strachwitz Claudius Schiller Kralik Iffland Sokrates
Katharina II. von Rußland Bellamy Schilling
Gerstäcker Raabe Gibbon Tschechow
Löns Hesse Hoffmann Gogol Wilde Gleim Vulpius
Luther Heym Hofmannsthal Morgenstern
Roth Klee Hölty Goedicke
Heyse Klopstock Kleist
Luxemburg Puschkin Homer Mörike
La Roche Horaz Musil
Machiavelli
Navarra Aurel Musset Kierkegaard Kraft Kraus
Lamprecht Kind Kirchhoff Hugo Moltke
Nestroy Marie de France
Laotse Ipsen Liebknecht
Nietzsche Nansen
Marx Ringelnatz
von Ossietzky Lassalle Gorki Klett Leibniz
May
vom Stein Lawrence Irving
Petalozzi
Platon Knigge
Sachs Pückler Michelangelo Kock Kafka
Poe Liebermann
de Sade Praetorius Mistral Zetkin Korolenko

The publishing house tredition has created the series **TREDITION CLASSICS.** It contains classical literature works from over two thousand years. Most of these titles have been out of print and off the bookstore shelves for decades.

The book series is intended to preserve the cultural legacy and to promote the timeless works of classical literature. As a reader of a **TREDITION CLASSICS** book, the reader supports the mission to save many of the amazing works of world literature from oblivion.

The symbol of **TREDITION CLASSICS** is Johannes Gutenberg (1400 – 1468), the inventor of movable type printing.

With the series, tredition intends to make thousands of international literature classics available in printed format again – worldwide.

All books are available at book retailers worldwide in paperback and in hardcover. For more information please visit: www.tredition.com

tredition®

tredition was established in 2006 by Sandra Latusseck and Soenke Schulz. Based in Hamburg, Germany, tredition offers publishing solutions to authors and publishing houses, combined with worldwide distribution of printed and digital book content. tredition is uniquely positioned to enable authors and publishing houses to create books on their own terms and without conventional manufacturing risks.

For more information please visit: www.tredition.com

The Food of the Gods A Popular Account of Cocoa

Brandon Head

Imprint

This book is part of the TREDITION CLASSICS series.

Author: Brandon Head
Cover design: toepferschumann, Berlin (Germany)

Publisher: tredition GmbH, Hamburg (Germany)
ISBN: 978-3-8495-1656-7

www.tredition.com
www.tredition.de

EAST INDIAN COOLIES ON A TRINIDAD CACAO ESTATE

CONTENTS.

LIST OF ILLUSTRATIONS AND MAPS.

Ceylon: A Hill Cacao Estate.

"THE FOOD OF THE GODS."

I. ITS NATURE.

When one thinks of the marvellously nourishing and stimulating virtue of cocoa, and of the exquisite and irresistible dainties prepared from it, one cannot wonder that the great Linnæus should have named it *theo broma*, "the food of the gods." No other natural product, with the exception of milk, can be said to serve equally well as food or drink, or to possess nourishing and stimulating properties in such well-adjusted proportions. Few, however, realize that in its stimulating properties cocoa ranks ahead of coffee, though below tea. As a matter of fact, the active principles of all three are alkaloids, practically identical and equally effective.[1] Each derives its value from its influence on the nervous system, which it stimulates, while checking the waste of tissue, but the cocoa-bean provides in addition solid food to replace wasted tissue. It is, indeed, so closely allied in composition to pure dried milk, that in this respect there is little to choose between an absolutely pure cocoa essence and the natural fluid.[2] It is this which makes it invaluable as an alternative food for invalids or infants.

Cacao Trees, Trinidad.

An early English writer on this valuable product spoke truly when he remarked: "All the American travellers have written such panegyricks, that I should degrade this royal liquor if I should offer any; yet several of these curious travellers and physicians do agree in this, that the cocoa has a wonderful faculty of quenching thirst, allaying hectick heats, of nourishing and fattening the body."

A modern writer[3] affords the same testimony in a more practical form when he records that: "Cocoa is of domestic drinks the most alimentary; it is without any exception the cheapest food that we can conceive, as it may be literally termed meat and drink, and were our half-starved artisans and over-worked factory children induced to drink it, instead of the in-nutritious beverage called tea, its nutritive qualities would soon develop themselves in their improved looks and more robust condition."

Such a drink well deserved the treatment it received at the hands of the Mexicans to whom we are indebted for it. At the royal banquets frothing chocolate was served in golden goblets with finely wrought golden or tortoise-shell spoons. The froth in this case was of the consistency of honey, so that when eaten cold it would gradually dissolve in the mouth. Here is a luscious suggestion for twentieth century housewives, handed to them from five hundred years ago!

AN-
CIENT MEXICAN DRINKING CUPS.
(*British Museum.*)

In health or sickness, infancy or age, at home or on our travels, nothing is so generally useful, so sustaining and invigorating. Far better than the majority of vaunted substitutes for human milk as an infant's food, to supplement what other milk may be available; incomparable as a family drink for breakfast or supper, when both tea and coffee are really out of place unless the latter is nearly all milk; prepared as chocolate to eat on journeys, and in many other ways, cocoa is a constant stand-by. Travelling in Eastern deserts on muleback, the present writer has never been without a tin of cocoa essence if he could help it, as, whatever straits he might be put to for provisions, so long as he had this and water, refreshment was possible, and whenever milk was available he had command in his lonely tent of a luxury unsurpassed in Paris or London. For the sustenance of invalids he has found nothing better in the home-land than a nightly cup of cocoa essence boiled with milk.

MOLINILLO (LITTLE MILL) OR CHOCOLATE WHISK.

Add to these experiences a love for the flavour which dates from childhood, and his admiration for this "food of the gods" will be appreciated, even if not sympathized in, by the few who have escaped its spell. Its value in the eyes of practical as well as scientific men is sufficiently demonstrated by its increasing use in naval and military commissariats, in hospitals, and in public institutions of all classes. In the British Navy, which down to 1830 consumed more cocoa than the rest of the nation together, it is served out daily, and in the army twice or thrice a week. Brillat Savarin, the author of the "Physiologie du Goût," remarks: "The persons who habitually take chocolate are those who enjoy the most equable and constant health, and are least liable to a multitude of illnesses which spoil the enjoyment of life."

A Cacao Harvest, Trinidad.

It certainly behoves us, therefore, to learn something more of such a valuable article than may be gleaned from the perusal of an advertisement, or the instructions on a packet containing it. There is something more than usually fascinating even in its history, in all the tales regarding this treasure-trove of the New World, and in the curious methods by which it has been treated. The story of its discovery takes us into the atmosphere of the Elizabethan period, and into the company of Cortes and Columbus; to learn of its cultivation and preparation we are transported to the glorious realms of the tropics, and to some of the most healthful centres of labour in the old country—in one case to the model village of the English Midlands. It is therefore an exceedingly pleasant round that lies before us in investigating this subject, as well as one which will afford much useful knowledge for every-day life.

Before proceeding to a closer acquaintance with the origin of cocoa, it may be well to clear the ground of possible misconceptions which occasionally cause confusion.

THE COCO-NUT PALM.

First, there is the word "cocoa" itself, an unfortunate inversion of the name of the tree from which it is derived, the cacao.[4] A still more unfortunate corruption is that of "coco-nut" to "cocoa-nut," which is altogether inexcusable. In this case it is therefore quite correct to drop the concluding "a," as the coco-nut has nothing whatever to do with cocoa or the cacao, being the fruit of a palm[5] in every way distinct from it, as will be seen from the accompanying illustration.

COCO-DE-MER.

The name "coco" is also applied to another quite distinct fruit, the *coco-de-mer*, or "sea-coco," somewhat resembling a coco-nut in its pod, but weighing about 28 lbs., and likewise growing on a lofty tree; its habitat is the Seychelles Islands. Sometimes also, confusion arises between the cacao and the coca or cuca,[6]a small shrub like a blackthorn, also widely cultivated in Central America, from the leaves of which the powerful narcotic cocaine is extracted.

LEAVES AND FLOWER OF THE CUCA SHRUB.

In the second place, the name "cocoa," which is strictly applicable only to the pure ground nib or its concentrated essence, is sometimes unjustifiably applied to preparations of cocoa with starch, alkali, sugar, etc., which it would be more correct to describe as "chocolate powder," chocolate being admittedly a confection of cocoa with other substances and flavourings.

Gathering Cacao: Santa Cruz, Trinidad.

"Chocolate" is, therefore, a much wider term than "cocoa," embracing both the food and the drink prepared from the cacao, and is the Mexican name, *chocolatl*, slightly modified, having nothing to do with the word cacao, in Mexican *cacauatl*.[7] In the New World it was compounded of cacao, maize, and flavourings to which the Spaniards, on discovering it, added sugar, cinnamon, vanilla, and other ingredients, such as musk and ambergris, cloves and nutmegs, almonds and pistachios, anise, and even red peppers or chillies. "Sometimes," says a treatise on "The Natural History of Chocolate," "China [quinine] and assa [fœtida?]; and sometimes steel and rhubarb, may be added for young and green ladies."

In our own times it is unfortunately common to add potato-starch, arrowroot, etc., to the cocoa, and yet to sell it by the name of the pure article. Such preparations thicken in the cup, and are preferred by some under the mistaken impression that this is a sign of its containing more nutriment instead of less. Although not so wholesome, there could be no objection to these additions so long as the preparations were not labelled "cocoa," and were sold at a lower price.

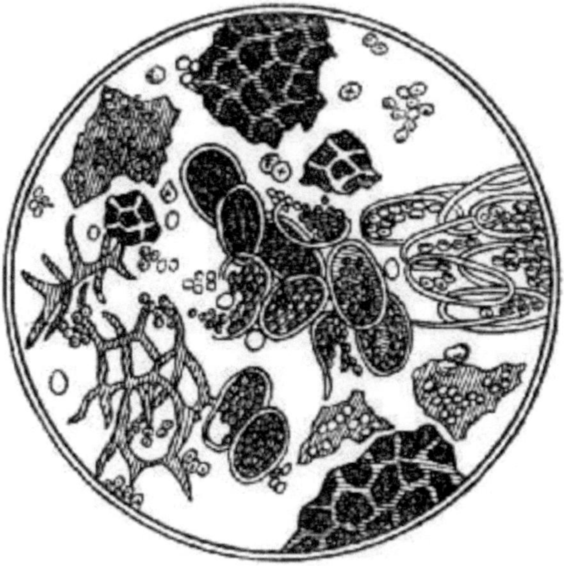

PURE DECORTICATED COCOA, HIGHLY MAGNIFIED.

Such adulteration is rendered possible by the presence in the bean of a large proportion of fatty matter or cocoa-butter, which renders it too rich for most digestions. To overcome this difficulty one or other of two methods is available: (1) Lowering the percentage of fat by the addition of starch, sugar, etc.; or (2) removing a large proportion of the fat by some extractive process; this latter method being in every respect preferable to that first mentioned.

COCOA ADULTERATED WITH ARROWROOT OR POTATO STARCH.

In order to avoid the expense and trouble consequent on the latter process, some manufacturers add alkali, by which means the free fatty acids are saponified, and the fat is held in a state of emulsion, thus giving the cocoa a false appearance of solubility.

Another effect of the alkali is to impart to the beverage a much darker colour, from its action on the natural red colouring matter of the cocoa, this darkening being often taken, unfortunately, as indicative of increased strength. On this account the presence of added alkali should be regarded as an adulteration, unless notified on the package in which the cocoa is contained.

A more subtle treatment with alkali for the same purpose is the addition to the pulverized bean of carbonate of ammonia, or caustic ammonia. This is afterwards volatilized by the application of heat. Scents and flavourings are then added to disguise their smell and taste.

Besides these combinations of cocoa with starch, sugar, etc., and cocoa treated with alkali, there are now found on the market mixtures of cocoa with such substances as kola, malt, hops, etc., sold under strange-sounding names, reminding one of the many mixtures that are made up as medicines rather than food. While the substances thus incorporated are of value in their place, they possess no virtues which are absent from the pure cocoa, and cannot be in any way considered an improvement of cocoa as food. The sooner this practice of drug taking under cover of diet comes to an end the better it will be for the national health.

Formerly Venetian red, umber, peroxide of iron, and even brick-dust, were employed to produce a cheaper article, but modern science and legislation combined have rendered such practices almost impossible. As early as the reign of George III. an Act[8] was passed, providing that, "if any article made to resemble cocoa shall be found in the possession of any dealer, under the name of 'American cocoa' or 'English cocoa,' or any other name of cocoa, it shall be forfeited, and the dealer shall forfeit £100." Yet this Act was allowed to become so much a dead letter that in 1851 the *Lancet* published the analysis of fifty-six preparations sold as "cocoa," of which only eight were free from adulteration. In some of the "soluble cocoas," the adulteration was as high as 65 per cent., potato starch in one case forming 50 per cent. of the sample. The majority of the samples were found to be coloured with mineral or earthy pigments, and specimens treated with red lead are on exhibition at South Kensington.

The inclusion of the husk or shell in some of the cheaper forms of chocolate is another reprehensible practice (strongly condemned), as they do not possess the qualities for which the kernel or nib is so highly prized. To prevent this practice it was enacted in 1770 that the shells or husks should be seized or destroyed, and the officer seizing them rewarded up to 20s. per hundredweight. From these a light, but not unpalatable, table decoction is still prepared in Ireland and elsewhere, under the designation of "miserables."

Among other beverages which have from time to time been produced from the cacao was a fermented drink much in vogue at the Mexican Court, to which it appears from the accounts of the conquest that Montezuma was addicted, as "after the hot dishes (300 in number) had been removed, every now and then was handed to him a golden pitcher filled with a kind of liquor made from cacao, which is very exciting." One variety, called *zaca*, drunk by the Itzas, consisted of cocoa mixed with a fermented liquor prepared from maize; but a more harmless invention was a drink composed of cocoa-butter and maize.

How the Cacao Grows.
(Showing Leaf, Flower, and Fruit.)

There remain three forms in which pure cocoa may be prepared as a beverage:

1. *Cocoa-nibs.*—The natural broken segments of the roasted cocoa-bean, after the shell has been removed, prepared for table as an infusion by prolonged simmering.

It is strange that this ridiculous and wasteful means is still in use at all, as next to none of the valuable portions of the nib are extracted. The quantity of matter removed by the hot water is so small, that close upon 90 per cent, of the nourishing and feeding constituents are left behind in the undissolved sediment, the substances extracted being principally salts and colouring matters. One can but suppose that the long habit of drinking an infusion from coffee-beans and tea-leaves has fixed in the mind the erroneous idea that the substance of the cocoa-bean is also valueless. The fact remains, however, that it is still customary at some hydropathic establishments, and perhaps in a few other instances, for doctors to order "nibs" for their patient, which may sometimes be accounted for by injury having resulted from drinking one of the many "faked" cocoas offered for sale; the order for "nibs" being a despairing effort to obtain the genuine article.

2. *Consolidated Nibs—i.e.,* cocoa-nibs ground between heated stones, whence it flows in a paste of the consistency of cream, which, when cool, hardens into a cake containing all the cocoa-butter. Cocoa in this form (mixed with sugar before cooling) is served in the British Navy—a somewhat wasteful and inconvenient practice, as when stirred, the excess of fat at once floats to the top of the cup, and is generally removed with a spoon, to make the drink more appetising.

3. *Cocoa Essence.*—This is the same article as No. 2, with about 60 per cent, of the natural butter removed; consequently the proportion of albuminous and stimulating elements is greatly increased. It is prepared instantly by pouring boiling water upon it, thus forming a light beverage with all the strength and flesh-forming constituents of the decorticated bean.[9]

Chemical analysis of cacao-nibs and cocoa essence shows them to contain on an average:

	Cacao-nibs.	Cocoa Essence.

Cocoa-butter	50	parts.	30	parts.
Albuminoid substances	16	"	22	"
Carbohydrates (sugar, starch, and digestible cellulose)	21	"	30	"
Theobromine	1.5	"	2	"
Salts	3.5	"	5	"
Other constituents	8	"	11	"
	---		---	
	100		100	

The *cocoa-butter* when clarified is of a pale yellow colour, and as it melts at about 90° F. it is of great value for pharmaceutical purposes, especially as it only becomes rancid when subjected to excessive heat and light, as to the direct rays of the sun.

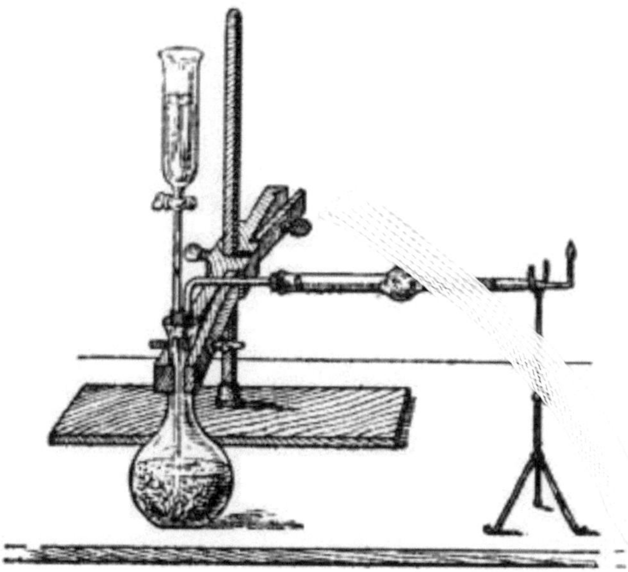

NA-

LYTICAL APPARATUS.

The *albuminoid* or *nitrogenous constituents* will be seen to form about a sixth of the whole nib, or more than a fifth of the cocoa essence, and to their presence is due the fact that absolutely pure cocoa is such a remarkable flesh-former.

Cacao Crop, Trinidad.

The *carbohydrates,* producing warmth and fat, are also important food substances, the proportion of which, while forming about a fifth of the whole bean, rises to close upon a third of the essence.

Cocoa also contains a *volatile oil,* from which it derives its peculiar and delicious aroma.

Thus *nearly nine-tenths of the cacao-bean may be assimilated by the digestive organs,* while three-fourths of tea and coffee are thrown away as waste. For the same bulk, therefore, cocoa is said to yield thirteen times the nutriment of tea, and four and a half times that of coffee. Its value as a substitute for mother's milk has already been alluded to, but may well be emphasized by a quotation from a paper read before the Surgical Society of Ireland in 1877 by one of its Fellows, Mr. Faussett:

"Without presuming to pass any judgment on the many artificial substitutes which, on alleged chemical and scientific principles, have from time to time been pressed forward under the notice of the profession and the public to take the place of mother's milk, I beg to call attention to a very cheap and simple article which is easily procurable—viz., cocoa, and which, *when pure and deprived of an excess of fatty matter*, may safely be relied on, as cocoa in the natural state abounds in a number of valuable nutritious principles, in fact, in every material necessary for the growth, development, and sustenance of the body."

After giving some remarkable cases of children being restored from "the last stage of exhaustion" by its use, and "continued through the whole period of infancy," with the effect of their becoming fine, healthy children, he concluded by saying:

"I beg therefore respectfully to commend cocoa, as an article of infant's food, to the notice of my professional brethren, especially those who, holding office under the Poor Laws, have such large and extensive opportunities of testing its value."

As a beverage for mothers or nurses cocoa is recommended by Dr. Milner Fothergill, in his work on "The Food we Eat," in preference to porter, stout or ale, an opinion now becoming generally adopted. It may, therefore, be regarded as the indispensable, all-round nursery food, if not the constant stand-by of the family.

That it is as nutritious for old as well as young we have an interesting proof in the fact that the first Englishman born in Jamaica, Colonel Montague James, who lived to the age of 104, took scarcely any food but cocoa and chocolate for the last thirty years of his life. For athletes and all who desire the development of the muscular tissues, its use is most beneficial. Professor Cavill, in his celebrated swim from Southampton to Portsmouth, and his nearly successful attempt to swim across the English Channel, considered it to be the most concentrated and sustaining food he could use for that trying test of endurance.

In his "Treatise on Food and Dietetics," Dr. Pavy remarks that:

"Containing, as pure cocoa does, twice as much nitrogenous matter, and twenty-five times as much fatty matter as wheaten flour, with a notable quantity of starch, and an agreeable aroma to tempt the palate, it cannot be otherwise than a valuable alimentary material. It has been compared in this respect to milk. It conveniently furnishes a large amount of agreeable nourishment in a small bulk, and, taken with bread, will suffice, in the absence of any other food, to furnish a good repast."

Indeed, the value of cocoa as food for ordinary mortals as well as for mythical beings cannot be better summed up than in the words of Professor Lankester, Superintendent of the Food Collections at South Kensington, who declares:

"It can hardly be regarded as a substitute for tea and coffee; it is, in fact, a substitute for all other kinds of food, and when taken with some form of bread, little or nothing else need be added at a meal. The same may be said of chocolate."

CACAO PODS

FOOTNOTES:

[1] According to Drs. Playfair and Lankester:

Tea	contains	3	per cent.	theine.
Coffee	"	1¾	"	caffeine.
Cocoa	"	2	"	theobromine.

Probably the proportion of caffeine in coffee would be more correctly stated as 1¼ per cent. Theine and caffeine are identical, but theobromine ($C_7H_8N_4O_2$) differs from both in the greater proportion of nitrogen which it contains.

[2] Dr. Johnson's analysis:

	Flesh formers in each hundred parts.
Dried milk	35
Cocoa essence	34¾
Cocoa-nibs	23
Best French chocolates	11

[3] Mr. O.L. Symonds, "Commercial Products of the Vegetable Kingdom."

[4] The *Cacao theobroma*. There are several other varieties of cacao, but none of them produce the famous food.

[5] The *Cocos nucifera*, or "nut-bearing coco."

[6]*Erythroxylon coca.*

[7] Or, as otherwise written, *cacava quahuitl.*

[8] 10 George III., c. 10.

[9] To make cocoa in perfection, for three breakfast-cups: in a quart jug (with rounded bottom and narrower neck by preference) mix 1½ dessert spoonfuls (¾ oz.) of Cocoa Essence with equal bulk of powdered white sugar, and stir to a thin paste with a little boiling water. Mix in an enamelled saucepan one breakfast-cup of milk with two cups of water (cups to be about ¾ full), and boil with care. When on the boil, pour this over the contents of the jug, and whisk vigorously for a few seconds (see illustration, p. 1). Serve to table without delay. To make a richer drink, use equal parts of milk and water. To ensure the beverage being served as hot as possible, it is desirable to warm the jug before the cocoa is put into it. The effect of this method of preparation is to impart to the cocoa a more mellow taste, and to produce a deep froth on the surface, giving it a

most appetizing appearance. The thorough mixing to which the cocoa is subjected also materially lessens the amount of sediment in the bottom of the cup.

II. ITS GROWTH AND CULTIVATION.

Cocoa is now grown in many parts of the tropics, reference to which is made in another chapter. The conditions, however, do not greatly vary, and there are probably many lands in the tropical belt where it is yet unknown that possess soil well suited to its extended cultivation.

The cacao-tree grows wild in the forests of Central America, and varieties have been found also in Jamaica and other West Indian islands, and in South America. It does not thrive more than fifteen degrees north or south of the equator, and even within these limits it is not very successfully grown more than 600 feet above the sea-level; in many districts where sugar formerly monopolized the plains, it was supposed that cocoa needed an altitude of at least 200 feet, but experiments of planting on the old sugar estates and other low-lying places are generally successful where the soil is good, as in Trinidad, Cuba, and British Guiana. It has been found that the expense saved in roads, labour, and transit on the level has been very considerable in comparison with that incurred on some of the hill estates.

In appearance the cacao-tree is not greatly unlike one of our own orchard trees, and trained by the pruning knife it grows similar in shape to a well-kept apple tree, no very low boughs being left, so that a man on horseback can generally pass freely down the long glades. Left to nature, it will in good soil reach a height of over twenty feet, and its branches will extend for ten feet from the centre.

Ceylon: Nursery of Cacao Seedlings in Baskets of plaited Palm Leaf.

The best soil is that made by the decomposition of volcanic rock, so that it is a common sight to find areas strewn with large boulders turned into a cocoa plantation of great fertility; but the best trees of all lie along the *vegas* which intersect the hills, where the soil is deep, and the stream winding among the trees supplies natural irrigation. The tree also grows well in loams and the richer marls, but will not thrive on clay and other heavy soils.

The cacao is one of the tenderest of tropical growths, and will not flourish in any exposed position, for which reason large shade belts are left along exposed ridges and other parts of a hill estate, thus greatly reducing the total area under cultivation, in comparison with an estate of equal extent on the level plains, where no shade belts are necessary.

The beans are planted either "at stake,"—when three beans are put in round each stake, the one thriving best after the first year being left to mature,—or "from nursery," whence, after a few months' growth in bamboo or palm-leaf baskets, they are transplanted into the clearing.

The preparation of the land is the first and greatest expense; trees have to be felled, and bush cut down and spread over the land, so that the sun can quickly render it combustible. When all is clear, the cacao is put in among a "catch crop" of vegetables (the cassava, tania, pigeon-pea, and others), and frequently bananas, though, as taking more nutriment from the soil, they are sometimes objected to. But the seedling cacao needs a shade, and as it is some years before it comes into bearing, it is usual to plant the "catch crop" for the sake of a small return on the land, as well as to meet this need.

In Trinidad, at the same time that the cacao[10] is planted at about twelve feet centres, large forest trees are also planted at from fifty to sixty feet centres, to provide permanent shade. The tree most used for this purpose is the *Bois Immortelle* (*Erythrina umbrosa*); but others are also employed, and experiments are now being made on some estates to grow rubber as a shade tree. In recent clearings in Samoa, trees are left standing at intervals to serve this end.

Samoa: Cacao in its fourth Year.

In Grenada, British West Indies, and some other districts, shade is entirely dispensed with, and the trees are planted at about eight feet centres, thus forming a denser foliage. By this means at least 500 trees will be raised on an acre, against less than 300 in Trinidad, the result showing almost invariably a larger output from the Grenada estates. This practice is better suited to steep hillside plantations than to those in open valleys or on the plains.

The cacao leaves, at first a tender yellowish-brown, ultimately turn to a bright green, and attain a considerable size, often fourteen to eighteen inches in length, sometimes even larger. The tree is subject to scale insects, which attack the leaf, also to grubs, which quickly rot the limbs and trunks, this last being at one time a very serious pest in Ceylon. If left to Nature the trees are quickly covered lichen, moss, "vines," ferns, and innumerable parasitic growths, and the cost of keeping an estate free from all the natural enemies which would suck the strength of the tree and lessen the crop is very great.

Young Cultivation, with catch Crop of Bananas, Cassava, and Tania: Trinidad.

The cacao will bloom in its third year, but does not bear fruit till its fourth or fifth. The flower is small, out of all proportion to the size of the mature fruit. Little clusters of these tiny pink and yellow blossoms show in many places along the old wood of the tree, often from the upright trunk itself, and within a few inches of the ground; they are extremely delicate, and a planter will be satisfied if every third or fourth produces fruit. In dry weather or cold, or wind, the little pods only too quickly shrivel into black shells; but if the season be good they as quickly swell, till, in the course of three or four months, they develop into full grown pods from seven to twelve inches long. During the last month of ripening they are subject to the attack of a fresh group of enemies—squirrels, monkeys, rats, birds, deer, and others, some of them particularly annoying, as it is often found that when but a small hole has been made, and a bean or so extracted, the animal passes on to similarly attack another pod; such pods rot at once. Snakes generally abound in the cacao regions, and are never killed, being regarded as the planter's best friends, from their hostility to his animal foes. A boa will probably destroy more than the most zealous hunter's gun.

PODS OF CACAO THEOBROMA.

From its twelfth to its sixtieth year, or later, each tree will bear from fifty to a hundred and fifty pods, according to the season, each pod containing from thirty-six to forty-two beans. Eleven pods will produce about a pound of cured beans, and the average yield of a large estate will be, in some cases, four hundredweight per acre, in others, twice as much. The trees bear nearly all the year round, but only two harvests are gathered, the most abundant from November to January, known as the "Christmas crop," and a smaller picking about June, known as the "St. John's crop." The trees throw off their old leaves about the time of picking, or soon after; should the leaves change at any other time, the young flower and fruit will also probably wither.

Of the many varieties of the cacao, the best known are the *criollo, forastero,* and *calabacilla.* The *criollo* ("native") fruit is of average size, characterized by a "pinched" neck and a curving point. This is the best kind, though not the most productive; it is largely planted in Venezuela, Columbia and Ceylon, and produces a bean light in colour and delicate in flavour. The *forastero* ("foreign") pod is long and regular in shape, deeply furrowed, and generally of a rough surface. The *calabacilla* ("little calabash") is smooth and round, like the fruit after which it is named. All varieties are seen in bearing with red, yellow, purple, and sometimes green pods, the colour not being necessarily an indication of ripeness.

Varieties of the Cacao.

On breaking open the pod, the beans are seen clinging in a cluster round a central fibre, the whole embedded in a white sticky pulp, through which the red skin of the cacao-bean shows a delicate pink. The pulp has the taste of acetic acid, refreshing in a hot climate, but soon dries if exposed to the sun and air. The pod or husk is of a porous, woody nature, from a quarter to half an inch thick, which, when thrown aside on warm moist soil, rots in a day or two.

Much has been written of life on a cocoa estate; and all who have enjoyed the proverbial hospitality of a West Indian or Ceylon planter, highly praise the conditions of their life. The description of an estate in the northern hills of Trinidad will serve as an example. The other industry of this island is sugar, in cultivating which the coloured labourers work in the broiling sun, as near to the steaming lagoon as they may in safety venture. Later on in the season the long rows between the stifling canes have to be hoed; then, when the time of "crop" arrives, the huge mills in the *usine* are set in motion, and for the longest possible hours of daylight the workers are in the field, loading mule-cart or light railway with massive canes. In the yard around the crushing-mills the shouting drivers bring their mule-teams to the mouth of the hopper, and the canes are bundled into the crushing rollers with lightning speed. The mills run on into the night, and the hours of sleep are only those demanded by stern necessity, until the crop is safely reaped and the last load of canes reduced to shredded *megass* and dripping syrup.

But upon the cocoa estate there is lasting peace. From the railway on the plain we climb the long valley, our strong-boned mule or lithe Spanish horse taking the long slopes at a pleasant amble, standing to cool in the ford of the river we cross and re-cross, or plucking the young shoots of the graceful bamboos so often fringing our path. Villages and straggling cottages, with palm thatch and *adobe* walls, are passed, orange or bread-fruit shading the little garden, and perhaps a mango towering over all. The proprietor is still at work on the plantation, but his wife is preparing the evening meal, while the children, almost naked, play in the sunshine.

The Home of the Cacao.
(*One of Messrs. Cadburys' Estates, Maracas, Trinidad.*)

The cacao-trees of neighbouring planters come right down to the ditch by the roadside, and beneath dense foliage, on the long rows of stems hang the bright glowing pods. Above all towers the *bois immortelle,* called by the Spaniards *la madre del cacao,* "the mother of the cacao." In January or February the *immortelle* sheds its leaves and bursts into a crown of flame-coloured blossom. As we reach the shoulder of the hill, and look down on the cacao-filled hollow, with the *immortelle* above all, it is a sea of golden glory, an indescribably beautiful scene. Now we note at the roadside a plant of dragon's blood, and if we peer among the trees there is another just within sight; this, therefore, is the boundary of two estates. At an opening in the trees a boy slides aside the long bamboos which form the gateway, and a short canter along a grass track brings us to the open savanna or pasture around the homestead.

Here are grazing donkeys, mules, and cattle, while the chickens run under the shrubs for shelter, reminding one of home. The house is surrounded with crotons and other brilliant plants, beyond which is a rose garden, the special pride of the planter's wife. If the sun has gone down behind the western hills, the boys will come out and play cricket in the hour before sunset. These savannas are the beauty-spots of a country clothed in woodland from sea-shore to mountain-top.

Ortinola, Maracas, Trinidad.

Next morning we are awaked by a blast from a conch-shell. It is 6.30, and the mist still clings in the valley; the sun will not be over the hills for another hour or more, so in the cool we join the labourers on the mule-track to the higher land, and for a mile or more follow a stream into the heart of the estate. If it is crop-time, the men will carry a *goulet*—a hand of steel, mounted on a long bamboo—by the sharp edges of which the pods are cut from the higher branches without injury to the tree. Men and women all carry cutlasses, the one instrument needful for all work on the estate, serving not only for reaping the lower pods, but for pruning and weeding, or "cutlassing," as the process of clearing away the weed and brush is called.

GOULET AND WOODEN SPOON.

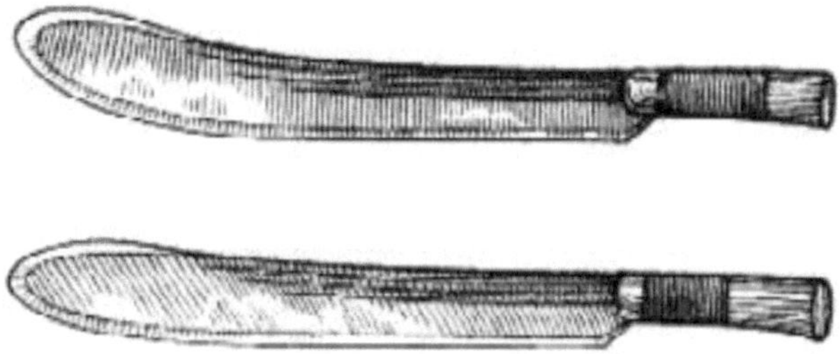

CUTLASSES.

Gathering the pods is heavy work, always undertaken by men. The pods are collected from beneath the trees and taken to a convenient heap, if possible near to a running stream, where the workers can refill their drinking-cups for the mid-day meal. Here women sit, with trays formed of the broad banana leaves, on which the beans are placed as they extract them from the pod with wooden spoons. The result of the day's work, placed in panniers on donkey-back, is "crooked" down to the cocoa-house, and that night remains in box-like bins, with perforated sides and bottom, covered in with banana leaves. Every twenty-four hours these bins are emptied into others, so that the contents are thoroughly mixed, the process being continued for four days or more, according to circumstances.

This is known as "sweating." Day by day the pulp becomes darker, as fermentation sets in, and the temperature is raised to about 140° F. During fermentation a dark sour liquid runs away from the sweat-boxes, which is, in fact, a very dilute acetic acid, but of no commercial value. During the process of "sweating" the cotyledons of the cocoa-bean, which are at first a purple colour and very compact in the skin, lose their brightness for a duller brown, and expand the skin, giving the bean a fuller shape. When dry, a properly cured bean should crush between the finger and thumb.

Cacao Drying in the Sun, Maracas, Trinidad.

Finally the beans are turned on to a tray to dry in the sun. They are still sticky, but of a brown, mahogany colour. Among them are pieces of fibre and other "trash," as well as small, undersized beans, or "balloons," as the nearly empty shell of an unformed bean is called. While a man shovels the beans into a heap, a group of women, with skirts kilted high, tread round the sides of the heap, separating the beans that still hold together. Then the beans are passed on to be spread in layers on trays in the full heat of the tropical sun, the temperature being upwards of 140° F.[11] When thus spread, the women can readily pick out the foreign matter and undersized beans. Two or three days will suffice to dry them, after which they are put in bags for the markets of the world, and will keep with but very slight loss of weight or aroma for a year or more.

Between crops the labourers are employed in "cutlassing," pruning, and cleaning the land and trees. Nearly all the work is in pleasant shade, and none of it harder than the duties of a market gardener in our own country; indeed, the work is less exacting, for daylight lasts at most but thirteen hours, limiting the time that a man can see in the forest: ten hours per day, with rests for meals, is the average time spent on the estate. Wages are paid once a month, and a whole holiday follows pay-day, when the stores in town are visited for needful supplies. Other holidays are not infrequent, and between crops the slacker days give ample time for the cultivation of private gardens.

Labourers from India are largely imported by the Government under contract with the planters, and the strictest regulations are observed in the matter of housing, medical aid, etc. At the expiration of the term of contract (about six years) a free pass is granted to return to India, if desired. Many, however, prefer to remain in their adopted home, and become planters themselves, or continue to labour on the smaller estates, which are generally worked by free labour, as the preparations for contracted labour are expensive, and can only be undertaken on a large scale.

Labourer's Cottage, Cacao Estate, Trinidad.
(Bread Fruit and Bananas.)

The natives of India work on very friendly terms with the coloured people of the islands, the descendants of the old African slaves, and the cocoa estate provides a healthy life for all, with a home amid surroundings of the most congenial kind.[12]

BASKETS OF CACAO ON PLANTAIN LEAVES.

In other cocoa-growing countries processes vary somewhat. On the larger estates artificial drying is slowly superseding the natural method, for though the sun at its best is all that is needed, a showery day will seriously interfere with the process, even though the sliding roof is promptly pulled across to keep the rain from the trays.

In Venezuela an old Spanish custom still prevails of sprinkling a fine red earth over the beans in the process of drying; this plan has little to recommend it, unless it be for the purpose of long storage in warehouses in the tropics, when the "claying" may protect the bean from mildew and preserve the aroma. In Ceylon it is usual to thoroughly wash the beans after the process of fermentation, thus removing all remains of the pulp, and rendering the shell more tender and brittle. Such beans arrive on the market in a more or less broken state, and it seems probable that they are more subject to contamination owing to the thinness of the shell. The best "estate" cocoa from Ceylon has a very bright, clear appearance, and commands a high price on the London market; this cocoa is of the pure *criollo* strain, light brown (pale burnt sienna) in colour.

CACAO TREE AND SEEDLING

The valleys of Trinidad and Grenada have grown cocoa for upwards of a hundred years, but up to the present time very little in the way of manuring has been done beyond the natural vegetable deposits of the forest. In many estates of recent years cattle have been quartered in temporary pens on the hills, moving on month by month, with a large central pen for the stock down on the savanna.

The cocoa-beans are shipped to Europe in bags containing from one to one and a half hundredweight, and are disposed of by the London brokers nearly every Tuesday in the year at a special sale in the Commercial Sale Room in Mincing Lane.

The cacao-tree has sometimes been grown from seed in hot-houses in this country, but always with difficulty, for not only must a mean temperature of at least 80° F. be maintained, but the tree must be shielded from all draught. Among the most successful are the trees grown by Mr. James Epps, Jun., of Norwood, by whose kind permission the accompanying sketches from life were made. Success has only crowned his efforts after many years of patient care. To grow a mere plant was comparatively simple, but to produce even a flower needed long tending, and involved much disappointment; while to secure fruition by cross-fertilization was a still more difficult task, accomplished in England probably on only one other occasion.

Bournville: "The Factory in a Garden."

FOOTNOTES:

[10] For full information on the subject of planting, see Simmond's "Tropical Agriculture" (Spon, London and New York); Nicholl's "Tropical Agriculture" (Macmillan).

[11] See plate facing p. 77.

[12] See *frontispiece.*

III. ITS MANUFACTURE.

Up to this point the operations described have taken place in the lands where cacao is produced. To watch the further processes in its development as an article of food, let us in imagination follow one of the shiploads of cacao on its sea journey from the far tropics to one of the countries of the old world, until the sacks of beans are finally deposited at a cocoa factory. An English factory, that of Messrs. Cadbury, at Bournville, affords an excellent illustration of its manufacture, not only because about a third of all the beans imported into this country are treated there, but also because this treatment is effected amid ideal surroundings. Half a century ago Messrs. Cadbury Brothers employed but a dozen or twenty hands, and until within the last twenty-six years the firm was established in the town of Birmingham. The need for greater accommodation for the rapidly growing business, and a desire to secure improved conditions for the work-people, led to the removal of the factory to a distance of about four miles south of the city. A number of cottages erected for the work-people in those early days became the nucleus of a great scheme which in the last few years has expanded into the model village of Bournville, a name taken from the neighbouring Bourn stream. Year by year the factory grew and developed, until the green hay-fields, with the trout stream flowing through them, became gradually covered with buildings. To-day the factory seems like a small town in itself, intersected by streets, and surrounded by its own railway. But the greenness of the country clings wherever a chance is afforded, ivy and other creepers adorning the brick walls, window boxes bright with flowers, and trees planted here and there; for no opportunity has been neglected of making the surroundings beautiful.

Bournville Cocoa Works: Office Buildings.

Taking train from the city, glimpses can be caught, as we near our destination, of the pretty houses and gardens of the village, forming a great contrast to the densely populated district of Stirchley on the other side of the line. Stepping on to the station, we are greeted by a whiff of the most delicious fragrance, which is quite enough of itself to betray the whereabouts of the great factory lying beneath us, of which from this point we have a fairly good bird's-eye view. Down the station steps, and a few yards up the lane to the left, with a playing field on one side, and on the other a plantation of fir-trees almost hiding the red brick and timbered gables of the office buildings, and we have arrived at the factory lodge. Looking through the open door down a vista of archways bowered in clematis and climbing roses, with an alpine rock garden at each side of the broad walk, we might almost imagine ourselves to be at the entrance to some botanical gardens. But a glance at the thousands of check hooks covering the inner wall of the lodge informs us that more than 2,400 girls pass in and out every day. The men's lodge is at a separate gate.

Before entering the works, a few steps further along the road will give us some idea of the many advantages gained by moving the factory out into the country. Just opposite the lodge a sloping path leads to the cycle-house, where some 200 machines are stored during work hours. Beyond this, in the middle of a flower garden, stands the Estate Office of the Bournville Village Trust, and in the background higher up a girls' pavilion can be seen through the trees. Behind it stretch asphalt tennis-courts and playing-fields, bordered by a belt of fine old trees, under whose shade wind pretty shrubbery walks lined with rustic seats. A passage under the road leads straight from the works into these beautiful grounds, and on a summer's day few prettier sights could be found than the numbers of white-robed girls who stream across in the dinner-hour to revel in the sunshine of the open fields, or sit in groups beneath the shady trees, enjoying a picnic lunch. A little further along the road the trees and the rhododendron bushes sweep backwards, leaving an open space, where a smooth lawn reaches to the front of a fine old mansion, for many years used as a home for some fifty of the work-girls whose own homes are at a distance, or who have no home at all. The fruit gardens and vineries belonging to "Bournville Hall" are used for the benefit of work-people who are ill.

Coronation Cricket Pavilion, Bournville.

Turning back again, we find on the other side of the road a magnificent pavilion, the Coronation gift of the firm to their employees, which overlooks the broad level stretch of one of the finest cricket grounds in the Midlands. Away in the hollow beyond, the Bourn forms a picturesque, shady pool, part of which is used to make a capital open-air swimming bath for the men. In the rising background are the pretty houses and the gardens of the model village. Still retracing our steps, we now come to the original cottages built by the firm. Plainer and less picturesque than those of more modern construction, their air of comfort, and the creepers which cover many of their walls, make them harmonize well with their surroundings. One of them is now used as a youths' club, providing games, a circulating library, and reading and lecture rooms. Another contains club rooms for the office staff. In passing we catch sight of a fine swimming bath for the girls.

Through the lodge and under the clematis, a few steps bring us to the private railway-station, which in size would do credit to many a town. Here trucks are loaded with finished goods and despatched to their various destinations. Every working day of the year a long train, extending often in the busiest season to as many as forty truck-loads, steams out of this station to scatter the productions of

Bournville over the face of the Earth. Close by the station we turn into the offices, where the fittings and general arrangement convey an air of refined solidity according well with the goods produced.

Girls' Dining Hall, Bournville.

Before proceeding to study the manufacture of cocoa essence and chocolate from the bean as it is imported, it will be interesting to see the careful provision that is made for the health and cleanliness of the workers, for in connection with any food nothing is of greater importance than the circumstances attending its preparation. A gratuitous sick club is provided by the firm for the employees, including the services of a doctor and three trained nurses. A special retiring room, comfortably furnished, is provided for girls needing a quiet hour's rest.

We are taken into the girls' dining-hall, capable of seating over two thousand at a time, fitted with benches, the backs of which are convertible into table tops. The far end of the dining-hall leads into the huge kitchen, to which the girls can bring their own dinners to be cooked, or where they can buy a large variety of things at coffee-house prices. Here again the health of the workers is carefully studied. Fruit is made a speciality, an experienced buyer being employed to insure its better supply. A private dining-room is provided for the forewomen.

Returning to the dining-hall, we descend a flight of steps into the spacious dressing-rooms, with vistas of wooden screens, filled on each side with numbered hooks. Here every morning the thousands of girls not only divest themselves of their outer garments, but change their dresses for washing frocks of white holland. The material for these is provided by the firm, free for the first, and afterwards at less than cost price, and the girls are required to start work in a clean frock every Monday morning. It will be seen at once how this helps them to keep neat and respectable; their strong white washing frocks only being soiled by their work, after which they change back into their own unstained clothes, and turn out looking as great a contrast to the usually pictured type of factory girl as can be imagined. The forewomen also conform to this arrangement, but wear washing dresses of blue cotton to distinguish them from the girls. Round the walls of this vast dressing-room hot-water pipes are placed, and over these are shelves where on a rainy day wet boots can be deposited to dry. Specially thoughtful is the provision of rubber snow-shoes, imported from America for their use, and supplied under cost price. Beneath each stool, too, is a shelf for heavy boots, which can be replaced in the factory by slippers.

Mention has already been made of the provision for illness or accidents, and of the care shown in the many arrangements for maintaining and improving the health and physical development of the girls. Further evidence of this is found in the airy and well-lighted work-rooms, from which funnels and exhaust fans collect and carry off all dust, and improve the ventilation, so that in spite of the multitudinous operations in progress, the whole place is kept as "spick and span" as a ship of the line. But another aggressive sign of the firm's belief in the motto *mens sana in corpore sano* is the presence of a lady whose whole time is devoted to the physical culture of the girls. Trained in Swedish athletics, this lady and her assistant undertake the teaching, not only of gymnastics, but of swimming and numerous games. Every day drill classes are held, an opportunity being thus provided for all the younger girls to attend a half-hour's lesson twice a week.

The result of all this thoughtful care is abundantly evident in the general air of health and comfort which pervades the whole factory, and in the bright faces which greet us at every turn, as we pass to and fro among the busy workers in this monster hive.

The Dinner Hour, Bournville.

Entering now, and turning into the private station, we see thousands of sacks of the freshly-imported beans being transferred to the neighbouring stores. The new arrivals must first be sifted and picked over to get rid of any that may be unsound, or of any foreign material still remaining. This is accomplished by a sorting and winnowing machine, which delivers by separate shoots the cleaned beans, graded according to size, and the dust and foreign matter.

A battery of roasters await the survivors of this operation, which are automatically conveyed to the hoppers. High-pressure steam supplies the requisite heat without waste or smoke, and as the huge drums slowly rotate, experienced workmen, on whose judgment great reliance is placed, carefully watch their contents, and decide when precisely the right degree of roasting has been attained to secure the richest aroma. Then they are passed through a cooling chamber, after which they are in condition for "breaking down."

This consists in cracking the shells of the beans, and releasing the kernels or "nibs," from which the shells and dust are winnowed by a powerful blast. It is accomplished by carrying the beans mechanically to the cracking machine at a considerable height, whence husks and nibs are allowed to fall before the winnower: the separated nibs are assorted according to size. Some of the shells find their way to the Emerald Isle, to be used by the peasants for the weak infusion called "miserables."

Now comes the important process of grinding, performed between horizontal mill-stones, the friction of which produces heat and melts the "butter," while it grinds the "nibs" till the whole mass flows, solidifying into a brittle cake when cold.

The thick fluid of the consistency of treacle flowing from the grinding-mills is poured into round metal pots, the top and bottom of which are lined with pads of felt, and these are, when filled, put under a powerful hydraulic press, which extracts a large percentage of the natural oil or butter. The pressure is at first light, but as soon as the oil begins to flow the remaining mass in the press-pot is stiffened into the nature of indiarubber, and upon this it is safe to place any pressure that is desired. As it is not advisable to extract all the butter possible, the pressure is regulated to give the required result. In the end a firm, dry cake is taken from the press, and when cool is ground again to the consistency of flour; this is the "cocoa essence" for which the firm of Cadbury is so well known in all parts of the world.[13]

Between cocoa and chocolate there are essential differences. Both are made from the cocoa nib, but whereas in cocoa the nibs are ground separately, and the butter extracted, in chocolate sugar and flavourings are added to the nib, and all are ground together into a paste, the sugar absorbing all the superfluous butter. If good quality cocoa is used, the butter contained in the nib is all that is needful to incorporate sugar and nib into one soft chocolate paste for grinding and moulding, but in the commoner chocolates extra cocoa butter has to be added. It is a regrettable fact that some unprincipled makers are tempted to use cheaper vegetable fats as substitutes for the natural butter, but none of these are really palatable or satisfactory in use, and none of the leading British firms are guilty of using such adulterants, or of the still more objectionable practice of grinding cocoa-shells and mixing them with their common chocolates.[14]

Flavouring is introduced according to the object in view; vanilla is largely employed in this country, though in France and Spain cinnamon is used, and elsewhere various spices. Willoughby, in his "Travels in Spain" (1664), writes:

"To every three and a half pounds of powder they add two pounds of sugar, twelve Vanillos, a little Guiny pepper (which is used by the Spaniards only), and a little Achiote[15] to give a colour. They melt the sugar, and then mingle all together, and work it up either in rolls or leaves."

Another writer says: "The usual proportion at Madrid to a hundred kernels of cocoa is to add two grains of Chile pepper, a handful of anise, as many flowers—called by the natives vinacaxtlides, or little ears—six white roses in powder, a pod of campeche,[16] two drachms of cinnamon, a dozen almonds and as many hazel-nuts, with achiote enough to give it a reddish tincture; the sugar and vanilla are mixed at discretion, as also the musk and ambergris. They frequently work this paste with orange water, which they think gives it a greater consistence and firmness."

Bournville Village: Laburnum Road.

When the chocolate is sufficiently ground it is put into a stove to attain the correct temperature, and is then passed on to a moulding-table, where it is pressed into tin moulds, and shaken till it settles. After passing through a refrigerating chamber, the contents of these moulds are ready as cakes of hard chocolate for putting up in the well-known blue "Mexican," or the dark-red "Milk," packets.

It would, of course, be interesting to proceed to an inspection of the many processes involved in making all the dainties that are prepared with chocolate, and of the numerous trades concerned in the production of packages, boxes, and fancy cases, did space permit. Room after room might be visited, bright in the daylight, or equally well lighted by electricity at night, humming with busy machines; some peopled with girls—among whom only men wearing a certain badge on their arms are allowed—some with men and boys, but all vibrating with a genial air of content as well as of busy occupation. Suffice it to say that half the handicrafts of the town seem represented in this centre of industry, in every department of which order and cheerfulness reign supreme. Each would require a chapter to do it justice, for everything employed in packing seems to be made on the premises, and that, too, on a system of piecework paid for, not at the lowest possible price, but on the basis of securing a satisfactory living wage to the average worker. No wonder the faces around are bright, no wonder that openings at the Bournville factory are in demand, and that long service for the firm is the boast of so many of the employees. Among these, a little band of about thirty still upholds the traditions of the old firm that laid the foundations of the present company in the city of Birmingham.

Packing Room, Bournville.

The work hours are forty-eight each week, and the wages depend both on age and length of service, no man of twenty-three years of age and over twelve months' service receiving less than 24s. weekly. There are no deductions for sick club or fines, the sick fund, as before stated, being a free gift from the company. Offences and late time are entered in a record book, and an opportunity is given to wipe off all past records by two years' good service. The Athletic Club, with over 500 voluntary subscribers, runs three cricket, four football, and two hockey teams, besides bowling, tennis, swimming, and other sports. One of the most interesting events of the Cricket Club is the annual match with a team representing Messrs. Fry and Sons, of Bristol, the oldest established cocoa firm in this country. In friendly opposition to the "Bournville Club" are the teams drawn from the "Youths' Club," and other outside organizations. A summer camp of over a hundred boys has been successfully held at the seaside for some years past.

The recent introduction of the system of suggestion-boxes throughout the works has been a great success. All employees are invited to make suggestions, which are dealt with each week by two committees, one for the men and one for the girls. Prizes amounting to about £80 are offered every half-year for the best suggestions. During the first seven months of operation over 1,000 suggestions were received, a very large percentage of which were found sufficiently useful to be adopted. The result has been to draw all sections closer together, as each feels sure of getting due credit for original ideas. Many important alterations in organization and methods of working have been carried into effect, entirely owing to this scheme.[17]

Bournville Village: Linden Road.

In order to encourage thrift (at the same time insuring privacy), a Savings Fund on a novel system has been working successfully for several years at Bournville. The fund was opened in Jubilee year by gifts of £1 to each employee who had been three years in the service of the firm, and 10s. to those employed for a shorter time. Deposits are received, and amounts withdrawn in the usual way during the year, through collectors in each department, the depositors' cards being called in quarterly for audit. At the end of each financial year, in May, interest at the rate of four per cent. is added to the amount standing to the credit of each depositor, and the whole amount paid over to the Post Office Savings Bank. At this time also, Post Office officials attend at the works, and enter the amounts to the credit of each depositor, issuing new Post Office Savings books where necessary. This system secures absolute privacy for the permanent savings, and places the fund upon a secure basis. As some evidence that the scheme is appreciated, it may be stated that the total balance transferred to the Post Office Savings Bank has averaged over £3,200 per annum.

While in the district of Bournville, the opportunity must not be lost of becoming more closely acquainted with the village around the works. Away beyond the factory stretches an estate of nearly

500 acres, set apart for the purpose of "alleviating the evils which arise from the insanitary and insufficient accommodation supplied to large numbers of the working classes, and of securing to workers in factories some of the advantages of outdoor village life, with opportunities for the natural and healthful occupation of cultivating the soil." As yet only some 450 houses have been erected, pretty, picturesque cottages all of them, for the most part semi-detached, each on its sixth of an acre, more or less, housing in all a population of about 2,000.

Fishing Pool, Bournville.

It was compassion for the ill-housed work-people of Birmingham that led Mr. George Cadbury, the founder of the village, to undertake so splendid a task, and having accomplished it, he crowned it by making a gift of the whole to the nation, placing its administration in the hands of a Trust. In doing so he laid down ideal stipulations for its development, and for the regulation of the villages which may in the future be built out of the income of the Trust. The principal of these are that factories or workshops shall never occupy more than one fifteenth of the area; that no house shall occupy more than one-fourth of the ground allotted to it; that in addition to wide roads and the ample gardens thus secured, one-tenth of the area shall be reserved for public open spaces for ever, parts of which are to be used as children's playgrounds. At present no intoxicants are sold or prepared on the estate, and if ever the trustees should see fit to permit this, it is to be as a co-operative undertaking, the profits of which shall "be devoted to securing for the village community recreation and counter-attraction to the liquor trade as ordinarily conducted."

Such a scheme affords a model for public bodies tackling the housing problem in earnest, and is fraught with great hopes for the future. The annual income, nearly £6,000, is to be applied first to the development of this estate, and subsequently to the purchase of estates near Birmingham or other large towns, and the establishment of new villages thereon. A most important feature is, that although the rents are calculated to yield a fair return on the cost, including a proportion of development expenses, they are so low that a five-roomed cottage with bath and every convenience can be had for the rent of a two-roomed hovel in the slums. About two-fifths of the householders find employment in the cocoa works, the rest in the adjoining villages or in Birmingham.

Almshouse Quadrangle, Bournville.

The gardens are a special feature, and before the houses are let, they are laid out by the Trust, and planted with fruit trees. All are well worked, and an average yield in vegetables and fruit of nearly two shillings a week has been found possible, equivalent to something like £60 an acre—more than twelve times as much food as would be produced if under pasturage. Two professional gardeners, with several men under them, are employed to look after the gardening department, and they are always ready to give any information or advice required by the tenants, so that the cottage gardens may be cultivated to the utmost profit. At present the public buildings consist of a village inn and baths; a school is shortly to be erected. Building is being steadily proceeded with, and although the development of the estate may be somewhat slow at first, it will advance with growing rapidity as the revenue increases. No wonder that there is an omnipresent air of comfort and prosperity, or that the death-rate is only about eight per thousand, in comparison with nineteen in the neighbouring city.

No description of Bournville would be complete without a mention of its picturesque alms-houses. Here a haven of rest is provided for some of those who, in their best years, have rendered faithful service to the firm. Thirty-three independent houses, brick and stone built, each with its own doorway to the quiet greensward, and its windows to the sun, form an inviting, reposeful quadrangle. They were the last gift of a life devoted to the interests of others, and the happiness and peace which characterize them are fitting memorials of the late Richard Cadbury, the elder of the two brothers who founded this great industry, and who have in their lives been favoured to see such untold blessing upon their labours.

SECTION OF A CHOCOLATE FACTORY.

The accompanying diagram of a chocolate factory is reproduced by kind permission of the Berlin publishers of Dr. Paul Zipperer's well-known work on "The Manufacture of Chocolate," which contains much valuable information. The machinery described is that of Messrs. Lehmann, of Dresden, one of the largest makers on the Continent.

By means of the lift (1) all the raw materials, sugar, cocoa, packing, etc., are carried up to the store-rooms (2). Here are the machines for cleansing and picking the raw cocoa-beans, which are fed into the elevator boxes (3) above the cleansing machine (4), which frees them from dust; they then pass to the continuous band (5) on which they are picked over, and from which they fall into movable boxes (6). They are thence transferred to the hoppers (7), and fed by opening a slide in the hopper, into the roasting machine (8). The quantity contained in the hoppers is sufficient to charge the roasting machine. When the roasting is completed the cocoa is emptied into trucks (9), and carried to the exhaust arrangement (10), where the beans are cooled down, the vapour given off passing out into the open air. At the same time the air of the roasting chamber is sucked out through the funnel-shaped tube fitted to the cover. The roasted cocoa is then passed to boxes (11), to be conveyed by the elevator to the crushing and cleansing machine (12). After being cleansed, the cocoa is carried in trucks (13) to hoppers (14) by which it is fed into the mills (15) on the lower floor. The sugar mill and sifting apparatus (26) placed near the crushing and cleansing machines are also fed by a hopper from above. Cocoa and sugar are now supplied to the mixing machine (16), to be worked together before passing to the rolls (17) by which the final grinding is effected. After passing once or more through the mill, the finished chocolate mass is taken to the hot-room (18), where it remains in boxes until further treated, after which it is taken to the moulding-room. In the mixer (19) the mass acquires the consistency and temperature requisite for moulding. The mass is then taken in lumps to the dividing machine (20), and cut into pieces of the desired size and weight. On the table (21) the moulds, lying upon boards, are filled with chocolate and then taken to the shaking-table (22). By means of a double lift (23) the moulded chocolate, still lying upon boards, is conveyed to the cooling-room or cellar, in which there are benches or frames (24) for receiving the moulds as they are slipped off the boards. The cellar has to be cooled artificially, according to situation. Adjoining the cellar is the wrapping-room (25), and further on the warehouse. The goods so far finished are then taken by the lift (1) to the rooms where they are packed for delivery.

FOOTNOTES:

[13] For ancient processes see Appendix I., p. 103.

[14] "Chocolate is an article so disguised in the manufacture that it is impossible to tell its purity or value. The only safeguard is to buy that which bears the name of a reputable maker."—Chambers, "Manual of Diet."

[15] The heart-leaved bixa, or anotta.

[16] Log-wood.

[17] The regulations adopted are so interesting that a place has been found for them in an Appendix (p. 106).

IV. ITS HISTORY.

[*From Dufour.*]

OLD DRAWING OF AN AMERICAN INDIAN, WITH CHOCOLATE-POT AND WHISK.

Although now cultivated in many other tropical countries, the cacao tree is one of the New World's rich gifts, first made known to our ancestors by the venturesome Spaniards, who probably became acquainted with its cultivation early in the sixteenth century, and spread the knowledge derived from the Mexicans and the inhabitants of Central America to their other colonies. They found cacao a more veritable mine of wealth than even the gold of which they procured such store. It is indeed a curious coincidence that in those countries of gold the cacao-beans were not only the form in which tribute was paid, but themselves passed as currency. On account of their use for this purpose by the Mexicans, Peter Martyr styled them *amygdalæ pecuniariæ*—"pecuniary almonds"—exclaiming: "Blessed money, which exempts its possessors from avarice, since it cannot be hoarded or hidden underground!"

Joseph Acosta tells us that "the Indians used no gold nor silver to trafficke in or buy withall ... and unto this day (1604) the custom continues amongst the Indians, as in the province of Mexico, instede of money they use cacao." The Aztecs also made use of cacao in this way, as many as 8,000 beans being legal tender—rather a task, one would imagine, for the money-changers.

Native Americans Preparing and Cooking Cocoa.
Ogibe's "America," 1671.

In Nicaragua this practice was so general that "none but the rich and noble could afford to drink it, as it was literally drinking money." A rabbit sold there for ten beans, "a tolerably good slave" for a hundred. Slaves must, however, have been at a discount just then, if the silver value of the beans was no greater than when Thomas Candish wrote in 1586: "These cacaos serve amongst them both for meat and money ... 150 of them being as good as a Real of Plate"—about 6d. "A bag," of unknown size, "was worth ten crowns." One of the storehouses of Montezuma, the last of the old independent Mexican Chieftains,[18] was found by the Spaniards to contain as much as 40,000 loads of this precious commodity, in wicker baskets which six men could not grasp.

John Ogilby, writing in 1671 of the produce of America, says:

"But much more beneficial is the cacao, with which Fruit New Spain drives a great Trade; nay, serves for Coin'd Money. When they deliver a Parcel of Cacao, they tell them by five, thirty, and a hundred. Their Charity to the Poor never exceeds above one Cacao-nut. The chief Reason for which this Fruit is so highly esteem'd, is for the Chocolate, which is made of the same, without which the Inhabitants (being so us'd to it) are not able to live. Before the Spaniards made themselves Masters of Mexico, no other Drink was esteem'd but that of the Cacao; none caring for Wine, notwithstanding the Soil produces Vines everywhere in great Abundance of itself."

From contemporary travellers' records are to be gleaned many such strange facts and stranger fancies regarding the precious bean and its products, some of them extremely quaint and curious. Bancroft, for instance, writing of the Maya races of the Pacific, tells us that "before planting the seed they held a festival in honour of their gods, Ekchuah, Chac, and Hobnil, who were their patron deities. To solemnize it, they all went to the plantation of one of their number, where they sacrificed a dog having a spot on its skin the colour of cacao. They burned incense to their idols, after which they gave to each of the officials a branch of the cacao plant." Palacio also tells us that "the Pipiles, before beginning to plant, gathered all seeds in small bowls, after performing certain rites with them before the idol, among which was the drawing of blood from different parts of the body with which to anoint the idol;" and, as Ximinez states, "the blood of slain fowls was sprinkled over the land to be sown."

[*From Bontekoe.*]
A CACAO PLANTATION.
(*One of the earliest illustrations of this subject known, showing the shade trees, and beans drying.*)

The idea that secret rites were necessary at the planting of cacao to counteract their ignorance of its requirements was long current also among the superstitious Spaniards, who similarly accounted for the early failures of the English, as witness the following amusing extract from a contribution to the *Harleian Miscellany* in 1690:

"Cocoa is now a commodity to be regarded in our colonies, though at first it was the principal invitation to the peopling of Jamaica, for those walks the Spaniards left behind them there, when we conquered it, produced such prodigious profit with so little trouble that Sir Thomas Modiford and several others set up their rests to grow wealthy therein, and fell to planting much of it, which the Spanish slaves had always foretold would never thrive, and so it happened: for, though it promised fair and throve finely for five or six years, yet still at that age, when so long hopes and cares had been wasted upon it, withered and died away by some unaccountable cause, though they imputed it to a black worm or grub, which they found clinging to its roots.... And did it not almost constantly die before, it would come into perfection in fifteen years' growth and last till thirty, thereby becoming the most profitable tree in the world, there having been £200 sterling made in one year of an acre of it. But the old trees, being gone by age and few new thriving, as the Spanish negroes foretold, little or none now is produced worthy the care and pains in planting and expecting it. Those slaves gave a superstitious reason for its not thriving, many religious rites being performed at its planting by the Spaniards, which their slaves were not permitted to see. But it is probable that, where a nation as they removed the art of making cochineal and curing vanilloes into their inland provinces, which were the commodities of those islands in the Indians' time, and forbade the opening of any mines in them for fear some maritime nation might be invited to the conquering of them, so they might, likewise, in their transplanting cocoa from the Caracas and Guatemala, conceal wilfully some secret in its planting from their slaves, lest it might teach them to set up for themselves by being able to produce a commodity of such excellent use for the support of man's life, with which alone and water some persons have been necessitated to live ten weeks together, without finding the least diminution of health or strength."

Grenada, B.W.I.: Samaritan Estate

(Showing trays which slide on rails; the iron covers slide over the whole in case of wet.)

However valuable this last quality rendered the newly-discovered drink, its method of preparation and the unwonted spices employed prevented its ready adoption abroad, although the Spaniards and Portuguese took to it more kindly than some of the northern races. Joseph Acosta, writing of Mexico and Peru, says:

"The cocoa is a fruite little less than almonds, yet more fatte, the which being roasted hath no ill taste. It is so much esteemed among the Indians (yea, among the Spaniards), that it is one of the richest and the greatest traffickes of New Spain. The chief use of this cocoa is in a drincke which they call chocholaté, whereof they make great account, foolishly and without reason: for it is loathsome to such as are not acquainted with it, having a skumme or frothe that is very unpleasant to taste, if they be not well conceited thereof. Yet it is a drincke very much esteemed among the Indians, whereof they feast noble men as they passe through their country. The Spaniards, both men and women, that are accustomed to the country, are very greedy of this chocholaté. They say they make diverse sortes of it, some hote, some colde, and put therein much of that chili: yea, they make paste thereof, the which they say is good for the stomacke, and against the catarre."

But this was not the only medicinal property attributed to "the food of the gods," for the Aztecs used to prescribe as a cure for diarrhœa and dysentery a potion prepared of cacao mixed with the ground bones of their giant ancestors, exhumed in the mountains. Such a very active principle was sure to make its enemies too, and several amusing attacks have survived to witness their own refutation. It was regarded by some as a violent inflamer of the passions, which should be prohibited to the monks; for, as one writer puts it, "if such an interdiction had existed, the scandal with which that holy order has been branded might have proved groundless." As late as 1712, after its use had become established in this country, the mentor of the *Spectator* writes: "I shall also advise my fair readers to be in a particular manner careful how they meddle with romances, chocolates, novels, and the like inflamers, which I look upon as very dangerous to be made use of during this great carnival" (the month of May).

MEXICAN DRINKING-VESSELS, ROLLING-PIN AND WHISK.

Some accounted for the assumed ill-effects of cocoa to its admixture with sugar in the form of chocolate, for a few years earlier a London doctor had declared that "coffee, chocolate, and tea were at the first used only as medicines while they continued unpleasant, but since they were made delicious with sugar they are become poison." Similarly, an anonymous assailant in a pamphlet "Printed at the Black Boy, over against St. Dunstan's Church, in Fleet Street," exclaims:

"As for the great quantity of sugar which is commonly put in, it may destroy the native and genuine temper of the chocolate, sugar being such a corrosive salt, and such an hypocritical enemy of the body. Simeon Pauli (a learned Dane) thinks sugar to be one cause of our English consumption, and Dr. Willis blames it as one of our universal scurvies: therefore, when chocolate produces any ill effects, they may be often imputed to the great superfluity of its sugar."

Cacao Tree, Trinidad.

In the New World fewer questions were raised, and the only conscientious objection appears to have been felt by a Bishop of Chiapa, whose performance of the Mass was disturbed by its use. The story is told in Gaze's "New Survey of the West Indies," published in 1648, and is worth repetition. It is well to bear in mind his information that "two or three hours after a good meal of three or four dishes of mutton, veal or beef, kid, turkeys or other fowles, our stomackes would bee ready to faint, and so wee were fain to support them with a cup of chocolatte."

"The women of that city, it seems, pretend much weakness and squeamishness of stomacke, which they say is so great that they are not able to continue in church while the mass is briefly hurried over, much lesse while a solemn high mass is sung and a sermon preached, unles they drinke a cup of hot chocolatte and eat a bit of sweetmeats to strengthen their stomackes. For this purpose it was much used by them to make their maids bring them to church, in the middle of mass or sermon, a cup of chocolatte, which could not be done to all without a great confusion and interrupting both mass and sermon. The Bishop, perceiving this abuse, and having given faire warning for the omitting of it, but all without amendment, thought fit to fix in writing upon the church dores an excommunication against all such as should presume at the time of service to eate or drinke within the church. This excommunication was taken by all, but especially by the gentlewomen, much to heart, who protested, if they might not eate or drinke in the church, they could not continue in it to hear what otherwise they were bound unto. But none of these reasons would move the Bishop. The women, seeing him so hard to be entreated, began to slight him with scornefull and reproachfull words: others slighted his excommunication, drinking in iniquity in the church, as the fish doth water, which caused one day such an uproar in the Cathedrall that many swordes were drawn against the Priests, who attempted to take away from the maids the cups of chocolatte which they brought unto their mistresses, who at last, seeing that neither faire nor foule means would prevail with the Bishop, resolved to forsake the Cathedrall: and so from that time most of the city betooke themselves to the Cloister Churches, where by the Nuns and Fryers they were not troubled....

"The Bishop fell dangerously sick. Physicians were sent for far and neere, who all with a joynt opinion agreed that the Bishop was poisoned. A gentlewoman, with whom I was well acquainted, was commonly censured to have prescribed such a cup of chocolatte to be ministered by the Page, which poisoned him who so rigorously had forbidden chocolatte to be drunk in the church. Myself heard this gentlewoman say that the women had no reason to grieve for him, and that she judged, he being such an enemy to chocolatte in the Church, that which he had drunk in his house had not agreed with his body. And it became afterwards a Proverbe in that country: 'Beware of the chocolatte of Chiapa!' ... that poisoning and wicked city, which truly deserves no better relation than what I have given of the simple Dons and the chocolatte-confectioning Doñas."

It was only natural that the nuns and friars of the cloister churches should raise no objection to this practice of chocolate drinking, for we read further that two of these cloisters were "talked off far and near, not for their religious practices, but for their skill in making drinkes which are used in those parts, the one called chocolatte, another atolle. Chocolatte is (also) made up in boxes, and sent not only to Mexico, but much of it yearly transported to Spain."

MODERN MEXICAN COCOA WHISK WITH LOOSE RINGS.
(*Brought home by the author.*)

The introduction of cocoa into Europe, indeed, as well as its cultivation for the European market, is due rather to the Jesuit missionaries than to the explorers of the Western Hemisphere. It was the monks, too, who about 1661 made it known in France. It is curious, therefore, to notice the contest that at one time raged among ecclesiastics as to whether it was lawful to make use of chocolate in Lent; whether it was to be regarded as food or drink. A consensus of opinion on the subject, published in Venice in 1748, states that

"Among the first Probabilist Theologians who undertook to write entire Treatises and to collect all the possible reasons as to whether the Indian beverage (chocolate) could agree with European fasting, was Father Tommaso Hurtado. He employed the whole of the Tenth Treatise of the second volume of the 'Moral Resolutions,' printed in 1651, and added thereto an Appendix of more chapters.

"Father Diana found reason for acquitting the consciences of those who, in time of fasting, should drink chocolate. Father Hurtado, more courageous withal, and more benign than Diana, does not speak of this treatise in order to investigate the law; the nature of fasting admits drinking without eating. Therefore consumers are, without the help of casuists, troubled themselves and afflicted, when in Lent they empty chocolate cups. Excited on the one hand by the pungent cravings of the throat to moisten it, reproved on the other by breaking their fast, they experience grave remorse of conscience; and, with consciences agitated and torn with drinking the sweet beverage, they sin. Under the guidance of these skilful theologians, the remorse aroused by natural and Divine light being blunted, Christians drink joyfully. For all agree that he will break his fast who eats any portion of chocolate, which, dissolved and well mixed with warm water, is not prejudicial to keeping a fast. This is a sufficiently marvellous presupposition. He who eats 4 ozs. of exquisite sturgeon roasted has broken his fast; if he has it dissolved and prepared in an extract of thick broth, he does not sin."

As for the introduction of cocoa into this country, the contemporary Gaze tells us that

"Our English and Hollanders make little use of it when they take a prize at sea, as, not knowing the secret virtue and quality of it for the good of the stomach, of whom I have heard the Spaniards say, when we have taken a good prize, a ship laden with cocoa, in anger and wrath we have hurled overboard this good commodity, not regarding the worth of it."

About the time of the Commonwealth, however, the new drink began to make its way among the English, and the *Public Advertiser* of 1657 contains the notice that "in Bishopsgate Street, in Queen's Head Alley, at a Frenchman's house, is an excellent West India drink, called chocolate, to be sold, where you may have it ready at any time, and also unmade, at reasonable rates." These rates appear to have been from 10s. to 15s. a pound, a price which made chocolate, rather than coffee, the beverage of the aristocracy, who flocked to the chocolate-houses soon to spring up in the fashionable centres. Here, records a Spanish visitor to London, were to be found such members of the polite world as were not at the same time members of either House. The chocolate-houses were thus the forerunners of our modern clubs, and one of them, "The Cocoa Tree," early the headquarters of the Jacobite party, became subsequently recognised as the club of the literati, including among its members such men as Garrick and Byron. White's Cocoa House, adjoining St. James' Palace, was even better known, eventually developing into the respectable White's Club, though at one time a great gambling centre.[19]

White's Club, on left of St. James's Palace.
(*From a Drawing of the time of Queen Anne.*)

A little later the "Indian Nectar," recommended by a learned doctor on account of "its secret virtue," was to be obtained of "an honest though poor man" in East Smithfield at 6s. 8d. a pound, or the "commoner sort at about half the price," so that it was getting within more general reach. Subsequently the following advertisement appeared regarding a patented preparation of cocoa "now sold at 4s. 9d. per pound."

"N.B.—The curious may be supplied with this superfine chocolate, that exceeds the finest sold by other makers, plain at 6s., with vanillos at 7s. To be sold for ready money only at Mr. Churchman's Chocolate Warehouse, at Mr. John Young's, in St. Paul's Churchyard, London, A.D. 1732."

The opportunities of increasing the revenue from the growing favourite were not lost sight of, and till 1820 its spread was checked by a duty of 1s. 6d. a pound, collected by the sale of stamped wrappers for each pound, half-pound, or quarter-pound, "neither more nor less," just as in the case of patent medicines at present.

In the reign of George III. the duty on colonial cocoa was raised to 1s. 10d. a pound, that on such as the East India Company imported to 2s., and that on all other sources of supply to 3s. In the early years of the last century the cocoa imported from any country not a British possession was charged no less than 5s. 10d. a pound as excise, with an extra Custom's duty of from 2½d. to 4¾d. on entry for home consumption. This restrictive tariff was by degrees relaxed, but it is only since 1853 that the duty has been reduced to 2d. a pound on the manufactured article, or 1d. a pound on the raw material.

While the heavy duties were in force, all houses in which the manufacture or sale of cocoa was carried on were compelled to have the fact stated over their doors, under penalty of £200 from the dealer having more than six pounds in his possession (who had to be licensed), and £100 from the customer encouraging the illicit trade. No less than £500 as fine and twelve months in the county gaol were inflicted for counterfeiting the stamp or selling chocolate without a stamp. To prevent evasion by selling the drink ready made, it was enacted under George I., whose physicians were extolling its medicinal virtues, that

"Notice shall be given by those who make chocolate for private families, and not for sale, three days before it is begun to be made, specifying the quantity, etc., and within three days after it is finished the person for whom it is made shall enter the whole quantity on oath, and have it duly stamped."

Nothing is more eloquent of the growing favour in which cocoa is held in this country, as its real value becomes more generally appreciated, than the remarkable progressive increase of the quantities imported during recent years, as will be seen from the table appended. These quantities doubled between 1880 and 1890, and have since more than doubled again.

TABLE SHOWING THE QUANTITIES OF CACAO CLEARED FOR HOME CONSUMPTION SINCE 1880.

	lbs.
1880	10,556,159
1881	10,897,795
1882	11,996,853
1883	12,868,170
1884	13,976,891
1885	14,595,168
1886	15,165,714
1887	15,873,698
1888	18,227,017
1889	18,464,164
1890	20,224,175
1891	21,599,860
1892	20,797,283
1893	20,874,995
1894	22,441,048
1895	24,484,502
1896	24,523,428
1897	27,852,152
1898	32,087,084
1899	34,013,812
1900	37,829,326
1901	42,353,724

1902	45,643,784

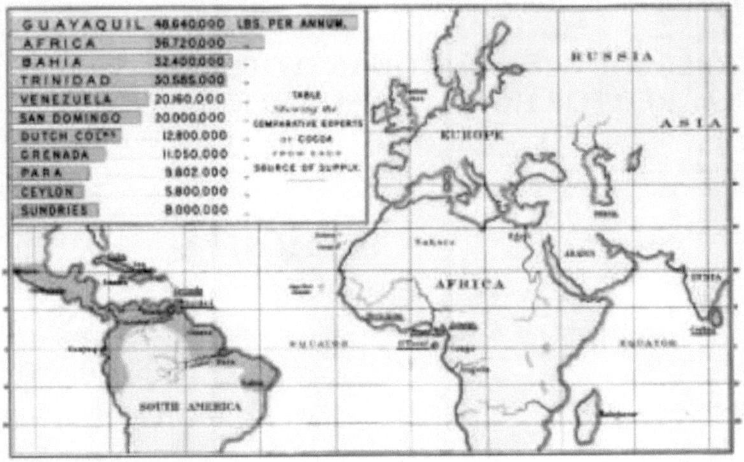

Map and Chart of Cocoa-Producing Countries

FOOTNOTES:

[18] Not an "Emperor," as reported by his conquerors.

[19] See Appendix III.

V. ITS SOURCES AND VARIETIES.

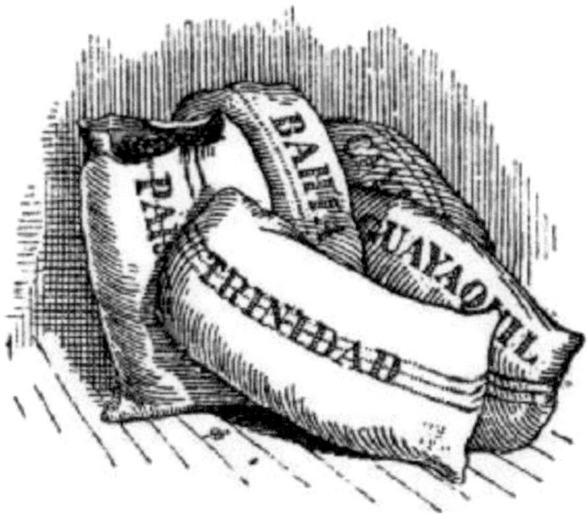

Guayaquil, in the republic of Ecuador, on the west coast of South America, produces the largest output in the world. This cacao has a bold bean and a fine flavour, and is rich in theobromine; it is much valued on the market, and its strength and character render it indispensable to the manufacturer.

The neighbouring countries of Columbia and Venezuela, facing the Caribbean Sea, have for centuries grown cacao of excellent quality. The *criollo* (creole) bean is generally used as seed, and for it high prices are obtained. Owing, however, to the unsettled state of the republics and their unstable governments, its cultivation has gone back rather than forward during the past decade. With better administration and settled peace, great developments might easily be achieved. The British Royal Mail Steam Packet Company provides a good fortnightly service to England.

In early times the Jesuit missionaries encouraged the natives to form small plantations on the borders of the river Orinoco, and Father Gumilla, in his "History of the Orinoco," says: "I have seen in these plains forests of wild cacao-trees, laden with bunches of pods, supplying food to an infinite multitude of monkeys, squirrels, parrots, and other animals."

The name of "Soconosco" cocoa is still a guarantee of excellent quality. This district in Guatemala was in bygone days so noted for its cacao that the whole crop was monopolized for the use of the Spanish Court. In Central America, as in other countries, the Spaniards gathered more solid riches from the cacao than from the gold mines they hoped to discover.

A Scene in the Maracas Valley, Trinidad.

British and Dutch Guiana produced but little cacao as long as sugar realized high prices, but in comparatively recent years it has been more extensively planted, and the crops from the lowlands at the mouths of the great South American rivers have been very heavy.

In French Guiana cacao was scarcely cultivated until about 1734, when a forest of it was discovered on a branch of the Yari, which flows into the Amazon. From this forest seeds were gathered, and plantations were laid out in Cayenne.

The cacao of Pará in Brazil differs from all other growths; the bean is much smaller and rounder, and is elongated, but when well cured it is mild, and has a very pleasant flavour, highly valued by manufacturers. Bahia produces large quantities of cacao, formerly of an inferior quality, owing to careless cultivation and indiscriminate mixing of all that was brought from the interior, some of it wild and uncured. But now this state of things is being improved, and the good quality of "fermented" Bahian cacao is fully recognised.

A little cacao is grown in the low-lying parts of Rio Janeiro, but it is not to be met with further south than this. The part of Florida which borders the Gulf of Mexico and the southern part of Louisiana mark the northerly limit of its natural growth.[20] A traveller in Louisiana in 1796 speaks of the cacao-tree among others as "covering with delightful shade the shores of the Mississippi," and on the banks of the Alatamaha in Georgia, but it is not cultivated so far north.

At the present day the West India Islands rival the South American Continent in providing cocoa from the New World. Trinidad has for more than a century deservedly claimed to be the first of these cocoa-producing islands. As far back as the sixteenth century the Spaniards who first colonized the island were interested in the cultivation of cacao. In the year 1780 a French gentleman residing in the neighbouring island of Grenada visited Trinidad, and gave such a glowing account of its fertility that agriculturists from France and elsewhere flocked to the colony, and ever since this date it has maintained a high standard of agricultural advance. The names of the cacao estates at the present day are nearly all Spanish or French, and throughout the British occupation of more than a hundred years the old families have in many cases held the same lands.[21]

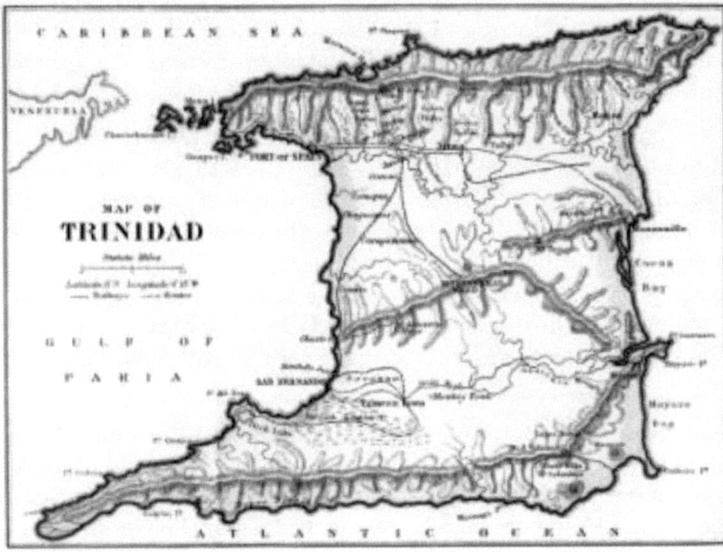

The oldest estates in the island lie in the northern valleys of Santz Cruz, Maracas, and Arima; but cultivation has been considerably extended in the Montserrat and Naparima districts, and more recently in almost every part of the island reached by the extension of the railway and the coasting steamboat. The Trinidad bean is the largest and finest flavoured, and commands a higher price on the market than any other from the West Indies.

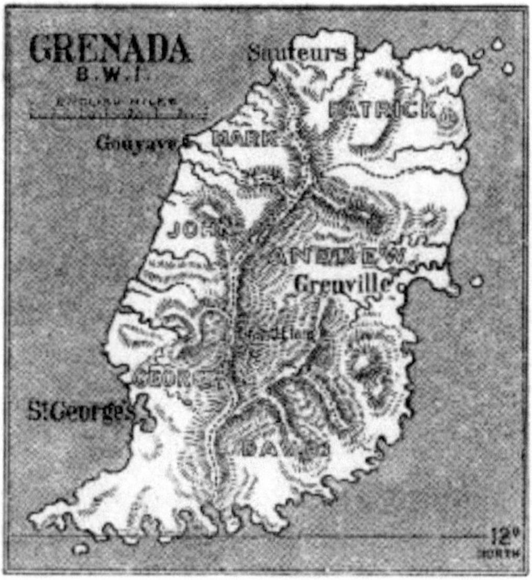

MAP OF GRENADA.

Next in importance to Trinidad is the little island of Grenada; here cacao is the staple industry, the sugar estates that once lined the shores having entirely disappeared. Grenada cacao is smaller than that of Trinidad, possibly on account of the different method of planting described in a previous chapter, but the flavour of the bean is exceedingly good and regular, and the crop is bought up eagerly on the British and American markets. The other West Indian islands producing cocoa are Jamaica and Dominica, where its cultivation is reviving; also St. Lucia, St. Vincent, Tobago, and Montserrat, each of which have a few plantations; those in St. Vincent suffered severely by the recent hurricane. The French islands of Guadeloupe and Martinique supply exclusively to the port of Havre; the cocoa from San Domingo is of a somewhat inferior quality. Cuba will probably considerably extend its output under American rule.

A Hill Cacao Estate, Grenada, B.W.I.

In the Eastern Hemisphere by far the largest supplies come from the small islands of St. Thomé and Principe, in the Gulf of Guinea, belonging to the Portuguese. These have in recent years proved especially adapted for the growth of the cacao, and the exports, especially from the island of St. Thomé, are very large; most of the crop finds its way to European markets, transhipping at Lisbon. There is little cacao grown in the mainland African colonies, though the German Government offers special inducements in the Kameruns; no British African colony grows it to any extent. Fernando Po sends supplies to Spain, and occasionally on the London market strange packages made of rough cowhide stitched with leather thongs are seen, containing beans from Madagascar.

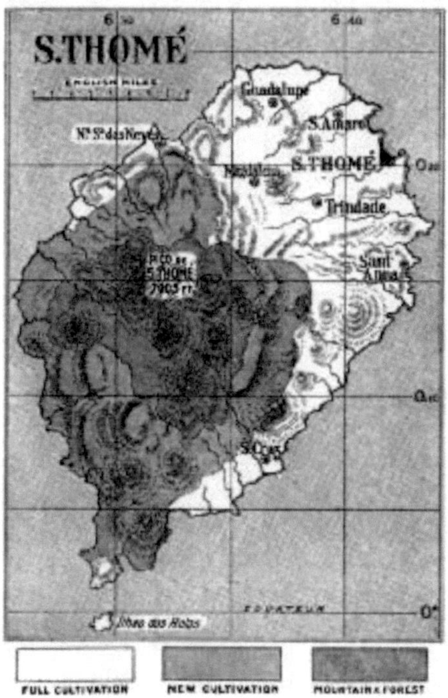

Ceylon: Carting Cacao to Rail.

MAP OF CEYLON.

Further east are the plantations of Ceylon. In the hill districts, of which Matale is the centre, are many estates, some in joint cultivation of tea and cocoa. The output from this colony is at the present time nearly stationary. The Dutch East Indian produce is almost exclusively shipped to Amsterdam.

In the preceding pages extracts have frequently been culled from writers of the past: in the literature of the present day Charles Kingsley's graphic account of Trinidad and its cacao and sugar plantations in "At Last" should be read *in extenso.* Another very interesting episode of modern date is the introduction of the cacao into the Samoan Islands in the Pacific by Robert Louis Stevenson. Writing to Sidney Colvin, on December 7, 1891, in one of his "Vailima Letters," he says:

"When I was filling baskets all Saturday, in my dull, mulish way, perhaps the slowest worker there, surely the most particular, and the only one that never looked up or knocked off, I could not but think I should have been sent on exhibition as an example to young literary men. 'Here is how to learn to write' might be the motto. You should have seen us; the veranda was like an Irish bog, our hands and faces were bedaubed with soil, and Faauma was supposed to have struck the right note when she remarked (*à propos* of nothing), 'Too much *eleele* (soil) for me.' The cacao, you must understand, has to be planted at first in baskets of plaited cocoa-leaf.[22] From four to ten natives were plaiting these in the wood-shed. Four boys were digging up soil and bringing it by the boxful to the veranda. Lloyd and I and Belle, and sometimes S. (who came to bear a hand), were filling the baskets, removing stones and lumps of clay; Austin and Faauma carried them when full to Fanny, who planted a seed in each, and then set them, packed close, in the corners of the veranda. From 12 on Friday till 5 p.m. on Saturday we planted the first 1,500, and more than 700 of a second lot. You cannot dream how filthy we were, and we were all properly tired."[23]

Samoa: A New Clearing for Cacao.

Three years later he records:

"I have been forbidden to work, and have been instead doing my two or three hours in the plantation every morning. I only wish somebody would pay me £10 a day for taking care of cacao, and I could leave literature to others."

Cacao cultivation in this island of Upolu has since that date developed wonderfully, and is attracting much attention, the first produce having been sold in Hamburg at a very high price. The consular report on Samoa published in February, 1903, states that "the mainstay of Samoa is cocoa," and it will be interesting to follow the progress of an industry of which the versatile Scotchman was an early pioneer.

FOOTNOTES:

[20] Florida even boasts a town of the name of Cocoa, but inquiries on the spot have failed to discover that any attempt was ever made to cultivate the plant there.

[21] Two of the coloured plates in this volume are reproductions of pictures by members of one of the oldest French families in the

island, painted on their cocoa estate in the beautiful valley of Santa Cruz.

[22] Leaf of the coco-nut palm.

[23] See plates facing pp. 27 and 29.

APPENDIX I.

ANCIENT MANUFACTURE OF COCOA.

Most of the operations described are only the performance on a large scale by modern machinery of those employed by the Mexicans, and by those who learned from them, of whom we read:

"For this purpose they have a broad, smooth stone, well polished or glazed very hard, and being made fit in all respects for their use, they grind the cacaos thereon very small, and when they have so done, they have another broad stone ready, under which they keep a gentle fire.

"A more speedy way for the making up of the cacao into chocolate is this: They have a mill made in the form of some kind of malt-mills, whose stones are firm and hard, which work by turning, and upon this mill are ground the cacaos grossly, and then between other stones they work that which is ground yet smaller, or else by beating it up in a mortar bring it into the usual form."

A later writer remarks of this process:

"The Indians, from whom we borrow it, are not very nice in doing it; they roast the kernels in earthen pots, then free them from their skins, and afterwards crush and grind them between two stones, and so form cakes of it with their hands."

A MEXICAN METATE, OR GRINDING STONE.

And, further on, in speaking of the Spaniards' mode of preparation, he says:

"They put them (the kernels) into a large mortar to reduce them to a gross powder, which they afterwards grind upon a stone. They make choice of a stone which naturally resists the fire, from sixteen to eighteen inches broad, and about twenty-seven or thirty long and three in thickness, and hollowed in the middle about one inch and a half deep. Under this they place a pan of coals to heat the stone, so that the heat makes it easy for the iron roller to make it so fine as to leave neither lump nor the least hardness."

At the present day, when the beans are plentiful on the cacao estates, but no machines for manufacture exist, the planters prepare a palatable drink by roasting the beans on a moving shovel or pan over the open fire, husking them by the time-honoured plan of tossing in the breeze, and grinding out on a flat stone in much the same manner as did the old Spaniards. The writer has even seen a little tobacco-press ingeniously adapted for the purpose of extracting the butter, the invention of Mr. J.H. Hart, of the Trinidad Botanical Gardens, a gentleman who has done much in the direction of investigating the best cacao for seed, and the most favourable methods of cultivation.

APPENDIX II.

BOURNVILLE WORKS SUGGESTION SCHEME.

OBJECTS.

December, 1902.

The objects in view are:

1. To encourage our employés to make all the suggestions they can for the mutual welfare of the business and everyone connected with it. Even the smallest suggestion may be of value.

2. To enable those in our employ to share in the benefit of the suggestions they make, and to receive personal recognition for them.

3. To insure harmonious relations between all sections of the work.

PRIZES.

Prizes of the undermentioned values will be given half-yearly for suggestions meriting reward:

MEN'S DEPARTMENTS.—One of £10; two of £5; two of £2 10s.; ten of £1; fifteen of 10s.; thirty of 5s. GIRLS' DEPARTMENTS.—One of £5; two of £2; eight of £1; fifteen of 10s.; thirty of 5s.

The following list will indicate on what lines suggestions may be made:

1. Comfort, safety, or health of employés.

2. Means by which waste of material may be avoided.

3. Saving of time or expense.

4. Improvements in machinery or in methods of working.

5. Introduction of new goods, or new ideas.

6. Calling attention to any existing defects.

7. Suggestions affecting athletic and other clubs and societies, libraries, magazine, etc.

8. Any suggestion not included in the above list will be welcomed.

REGULATIONS.

Everyone, including foremen and forewomen, is encouraged to make suggestions which, if of value, will be eligible for the prizes mentioned above (excepting those sent in by foremen and forewomen).

Suggestions should be written on or attached to the forms which will be found on each box, the boxes being fixed in the various departments, also in the entrance lodges, dining-rooms, and recreation grounds. Suggestions can be placed in any of these.

It is imperative that all particulars at head of form, which will bear a distinctive number, should be carefully filled in. If this is not complied with no notice will be taken of suggestions. Forms may be taken from the book and filled up at home.

All suggestions will be acknowledged by a notice posted on the boards once a week, giving a list of the printed numbers on the suggestion forms received for consideration.

Should any number not appear in this list a communication should at once be sent to the Secretary.

Those who have left the employ of the firm are entitled to prizes for any suggestions made whilst they were here, unless they should leave through misconduct.

The suggestions are considered weekly by the committees with a member of the firm, and are dealt with in the order in which they are received. They are finally judged by the firm at the end of May and November, and prizes distributed before the summer holidays and at the Christmas gathering.

Every effort is made by the committees to keep the names of the suggestors *strictly private.*

APPENDIX III.

THE EARLY COCOA HOUSES.

At No. 64, St. James's Street is the "Cocoa Tree Club." In the reign of Queen Anne there was a famous chocolate-house known as the "Cocoa Tree," a favourite sign to mark that new and fashionable beverage. Its frequenters were Tories of the strictest school. De Foe tells us in his "Journey through England," that "a Whig will no more go to the 'Cocoa Tree' ... than a Tory will be seen at the coffee-house of St. James's." In course of time the "Cocoa Tree" developed into a gaming-house and a club.

As a club, the "Cocoa Tree" did not cease to keep up its reputation for high play. Although the present establishment bearing the name dates its existence only from the year 1853, the old chocolate-house was probably converted into a club as far back as the middle of the last century. Lord Byron was a member of this club, and so was Gibbon, the historian.

—From "Old and New London," Cassell & Co.

NOTE.

Reference in detail to the numerous authorities who have been laid under contribution for this brochure would be out of place in so popular a compilation, but the writer desires to express his special indebtedness to "Cocoa: All about It" by "Historicas," not only for facts, but also for some of his illustrations. To Messrs. Cadbury, too, he is indebted for permission to use several of the illustrations, as well as for much valuable information.